SURVEYI

M

An

Jim Crume P.L.S., M.S., CFedS

Co-Authors

Cindy Crume

Bridget Crume

Troy Ray R.L.S.

Mark Sandwick L.S.I.T.

PRINTED EDITION

*****

PUBLISHED BY:

Jim Crume P.L.S., M.S., CFedS

Create Rectangular Coordinates

Book 2 of this Math-Series

First publication: November, 2013

Printed by CreateSpace

Available on Kindle and other devices

TERMS AND CONDITIONS

The content of the pages of this book is for your general information and use only. It is subject to change without notice.

Neither we nor any third parties provide any warranty or guarantee as to the accuracy, timeliness, performance, completeness or suitability of the information and materials found or offered in this book for any particular purpose. You acknowledge that such information and materials may contain inaccuracies or errors and we expressly exclude liability for any such inaccuracies or errors to the fullest extent permitted by law.

Your use of any information or materials in this book is entirely at your own risk, for which we shall not be liable. It shall be your own responsibility to ensure that any products, services or information available in this book meet your specific requirements.

Table of Contents

INTRODUCTION

Straight forward Step-by-Step instructions.

This book is just one part in a series of digital and printed editions on Surveying Mathematics Made Simple. The subject matter in this book will utilize the methods and formulas that are covered in the books that precede it. If you have not read the preceding books, you are encouraged to review a copy before proceeding forward with this book.

For a list of books in this series, please visit:

http://www.cc4w.net/ebooks.html

Prerequisites for this book: A basic knowledge of geometry, algebra and trigonometry is required for the explanations shown in this book.

Definition: Rectangular Coordinates (a.k.a. Cartesian Coordinates) is a coordinate system that specifies each point uniquely in a horizontal plane by a pair of numerical coordinates, which are the distances from the point of two fixed perpendicular directed lines, measured in the same unit of length. Each reference line is called a coordinate axis and the point where they meet is its origin, usually at an ordered pair of (0,0). The left-right axis is referred to as the 'X" axis and the up-down axis is referred to as the "Y" axis.

The X-Y coordinate reference system is used in mathematics. The N-E coordinate reference system is used in survey related professions. To understand the basis of the N-E coordinate system you must first understand the X-Y coordinate system.

X-Y COORDINATE SYSTEM

Figure 1 below illustrates the X-Y coordinate system.

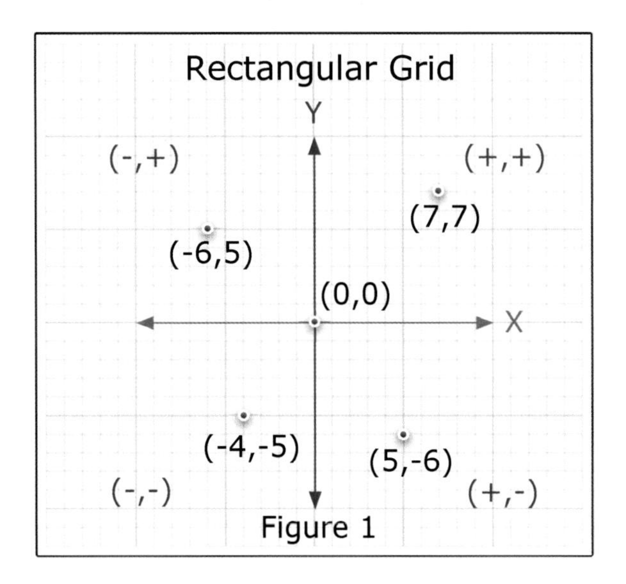

Figure 1

For the X-Y rectangular coordinate system, each paired coordinates are shown as (X,Y) values. The X coordinate is shown first and the Y coordinate is shown second separated by a comma. With the origin being (0,0) the distance in the X direction for the paired coordinate (-6,5) in the upper left quadrant is -6 units left of the Y axis and 5 units up from the X axis.

The rectangular grid is divided into equal units in both the X and Y direction with parallel lines extending up and down from the X axis and left and right from the Y axis. The intersection lines are perpendicular to each other.

Figure 2 below shows the relationship between polar coordinates to rectangular coordinates.

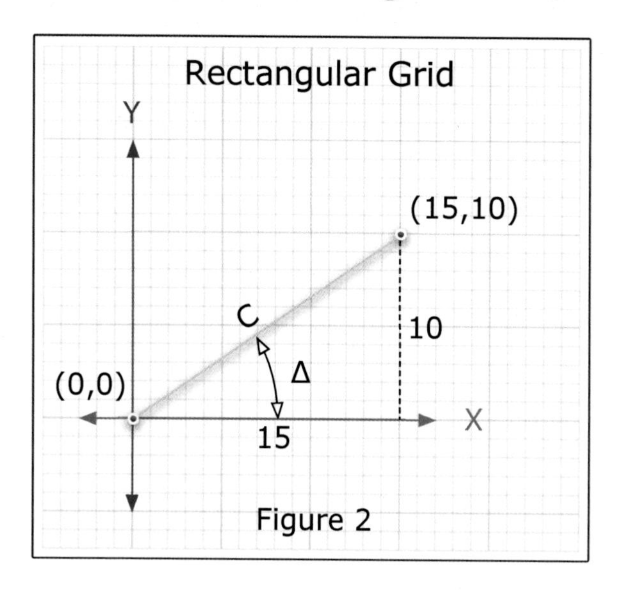

Figure 2

Line C and angle Δ are known as Polar Coordinates.

NOTES

N-E COORDINATE SYSTEM

Figure 3 below illustrates the N-E coordinate system.

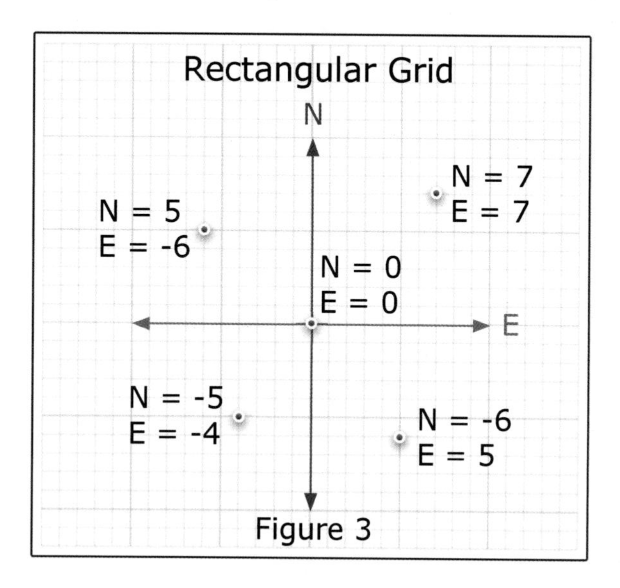

Figure 3

For the N-E rectangular coordinate system, each paired coordinates are shown as Northing and Easting values. The North coordinate is shown on top of the East coordinate. With the origin being N = 0 and E = 0 the distance in the North direction for the paired coordinate N = 5 and E = -6 in the upper left quadrant is 5 units north of the E axis and -6 units west from the N axis.

The rectangular grid is divided into equal units in both the N and E direction with parallel lines extending north and south from the E axis and east and west from the N axis. The intersection lines are perpendicular to each other.

Note: In the N-E coordinate system, you should avoid using negative coordinates. The origin N = 0 and E = 0 should always be far enough south and west so as to be out of your project area.

In the N-E coordinate system, polar coordinates will be synonymous with the bearing and distance for a line.

Example 1:

Figure 4 shows a line with a bearing and distance from a starting coordinate point of N = 10000 and E = 10000. When using assumed coordinates for a starting point, use a value that is to the nearest 10000 for the northing and easting. This makes it easy to look at the coordinate pair and tell if it is the starting point.

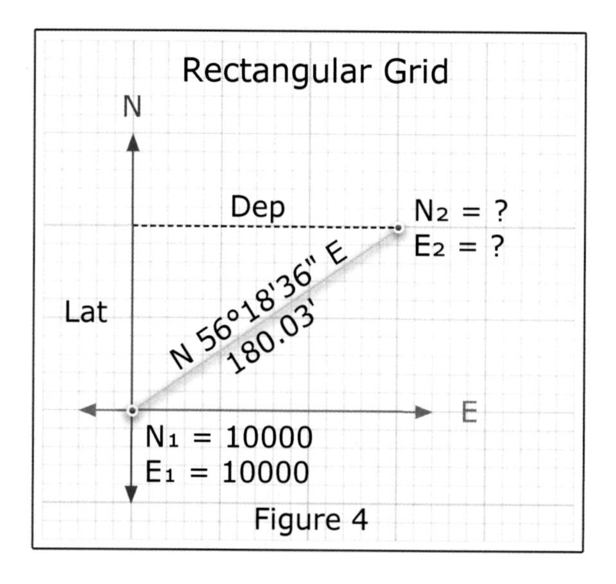

Figure 4

For the line N 56°18'36" E, 180.03' the starting coordinate point is N_1 = 10000 and E_1 = 10000. In order to calculate the ending coordinate point, the Latitude and Departure will need to be calculated for the line.

The Latitude (Lat) is the distance along the N-S axis from the starting point north coordinate to the ending point north coordinate of the line and the Departure (Dep) is the distance along the E-W axis from the starting point east coordinate to the ending point east coordinate of the line.

Use the following formulas to calculate the Latitude, Departure, Northing and Easting for the line above:

Formulas:

Lat (North-South direction) = Cos (Bearing*) x Distance

Dep (East-West direction) = Sin (Bearing*) x Distance

$N2 = N1 + Lat$

$E2 = E1 + Dep$

* Bearing angle must be in decimal degrees before getting the Cos or Sin value.

Note: See Book 1 "Bearings and Azimuths" of this math series for steps required to convert degrees-minutes-seconds to decimal degrees. All calculators require decimal degrees for trigonometric functions.

Example 1 - Figure 4 solution:

Lat = Cos (56°18'36") x 180.03' = **99.86250**

Dep = Sin (56°18'36") x 180.03' = **149.79413**

N2 = 10000 + 99.86250 = **10099.86250**

E2 = 10000 + 149.79413 = **10149.79413**

Example 2:

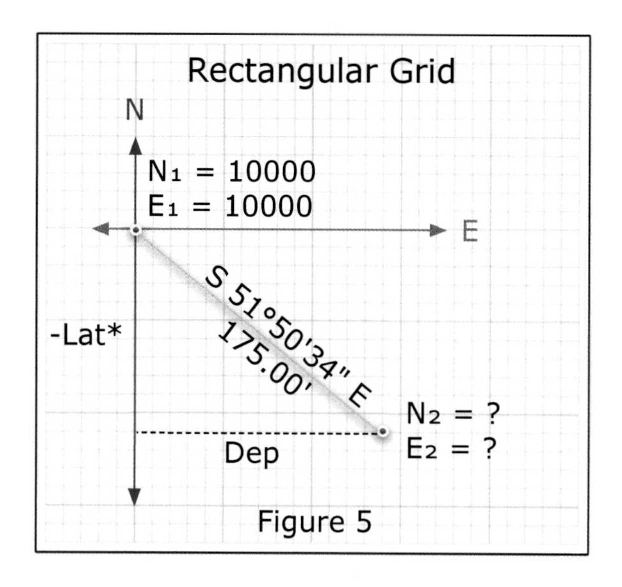

Figure 5

* Note the negative algebraic sign

Example 2 - Figure 5 solution:

Lat = Cos (51°50’34”) x 175.00’ = **-108.11876**

Dep = Sin (51°50’34”) x 175.00’ = **137.60572**

Note: The Lat is negative because the bearing is in a SE direction.

N2 = 10000 + (-)108.11876 = **9891.88124**

E2 = 10000 + 137.60572 = **10137.60572**

Examples 1 & 2 walked you through the steps to calculate the rectangular coordinates for the end point of a line. Now that you have a couple of examples to follow, try solving Examples 3 & 4, then review the solutions at the end of the book to see how you did.

Example 3:

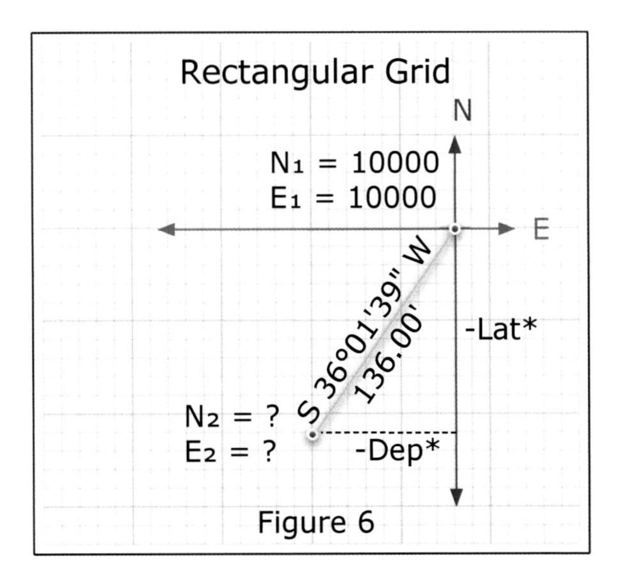

Figure 6

* Note the negative algebraic sign

Figure 6 solution:

Lat = Cos (??°??'??") x ? = ?

Dep = Sin (??°??'??") x ? = ?

N2 = 10000 + ? = ?

E2 = 10000 + ? = ?

The solution is towards the end of the book.

Example 4:

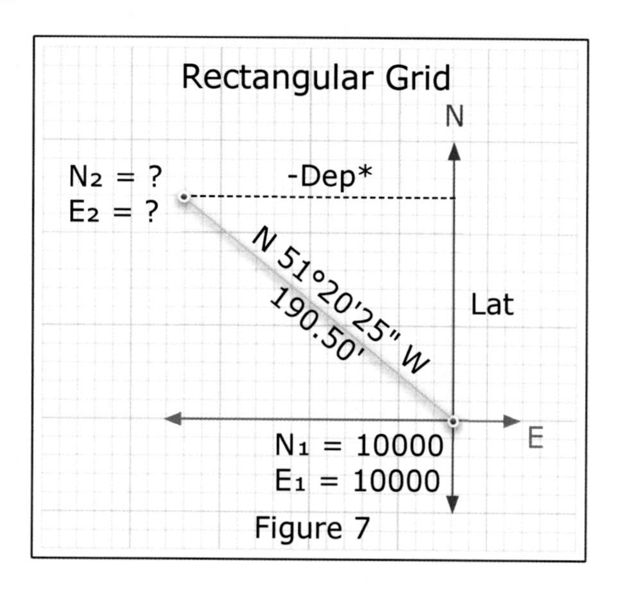

Figure 7

* Note the negative algebraic sign

Figure 7 solution:

Lat = Cos (??°??'??") x ? = ?

Dep = Sin (??°??'??") x ? = ?

N2 = 10000 + ? = ?

E2 = 10000 + ? = ?

The solution is towards the end of the book.

NOTES

SOLUTIONS TO EXAMPLES 3 & 4

Example 3 - Figure 6 solution:

Lat = Cos (36°01'39") x 136.00' = **-109.98793**

Dep = Sin (36°01'39") x 136.00' = **-79.99159**

Note: The Lat and Dep are negative because the bearing is in a SW direction.

N2 = 10000 + (-)109.98793 = **9890.01207**

E2 = 10000 + (-)79.99159 = **9920.00841**

Example 4 - Figure 7 solution:

Lat = Cos (51°20'25") x 190.50' = **119.00418**

Dep = Sin (51°20'25") x 190.50' = **-148.75569**

Note: The Dep is negative because the bearing is in a NW direction.

N2 = 10000 + 119.00418 = **10119.00418**

E2 = 10000 + (-)148.75569 = **9851.24431**

When performing trigonometric functions on a calculator using bearing angles, first you need to convert the angle to decimal degrees before using the Cos or Sin function on the calculator, multiply the Cos or Sin of the angle by the distance and finally change the sign to a negative depending on the direction of the line.

NE Bearing: Lat and Dep will be positive

SE Bearing: Lat will be negative and Dep will be positive

SW Bearing: Lat and Dep will both be negative

NW Bearing: Lat will be positive and Dep will be negative

Tip: By using North Azimuth (Naz) angles instead of Bearing angles for trigonometric operations, the calculator will automatically give you the correct sign for the Lat and Dep.

Example:

N56°18’36”E = **Naz 56°18’36”**

Cos (56°18’36”) = **0.554699**

Sin (56°18’36”) = **0.832051**

S51°50’34”E = **Naz 128°09’26”**

Cos (128°09’26”) = **-0.617821**

Sin (128°09’26”) = **0.786318**

S36°01'39"W = **Naz 216°01'39"**

Cos (216°01'39") = **-0.808734**

Sin (216°01'39") = **-0.588173**

N51°20'25"W = **Naz 308°39'35"**

Cos (308°39'35") = **0.624694**

Sin (308°39'35") = **-0.780870**

ABOUT THE AUTHOR

Jim Crume P.L.S., M.S., CFedS

My land surveying career began several decades ago while attending Albuquerque Technical Vocational Institute in New Mexico and has traversed many states such as Alaska, Arizona, Utah and Wyoming. I am a Professional Land Surveyor in Arizona, Utah and Wyoming. I am an appointed United States Mineral Surveyor and a Bureau of Land Management (BLM) Certified Federal Surveyor. I have many years of computer programming experience related to surveying.

This book is dedicated to the many individuals that have helped shape my career. Especially my wife Cindy. She has been my biggest supporter. She has been my instrument person, accountant, advisor and my best friend. Without her, I would not be the professional I am today. Cindy, thank you very much.

Other titles by this author:

http://www.cc4w.net/ebooks.html

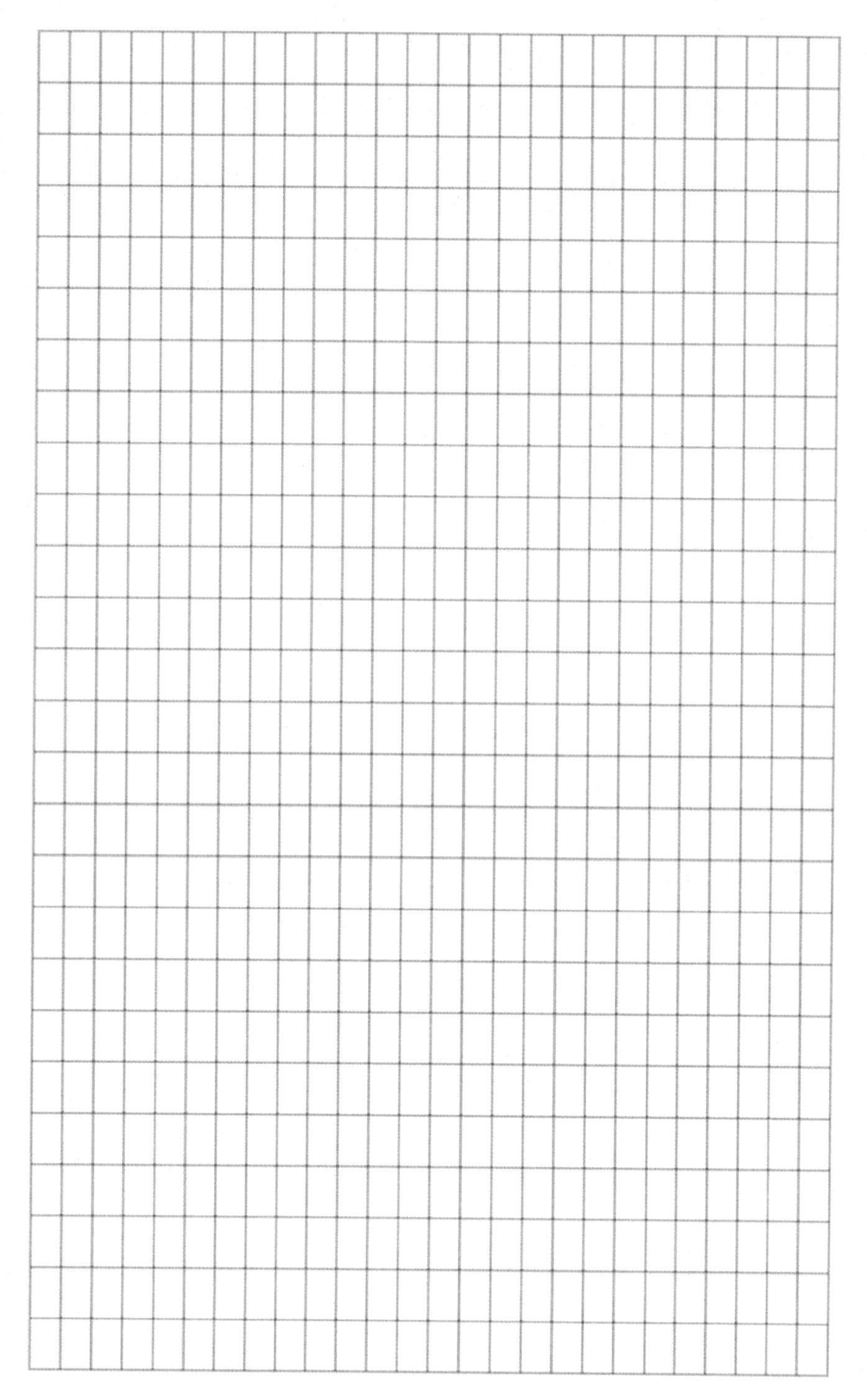

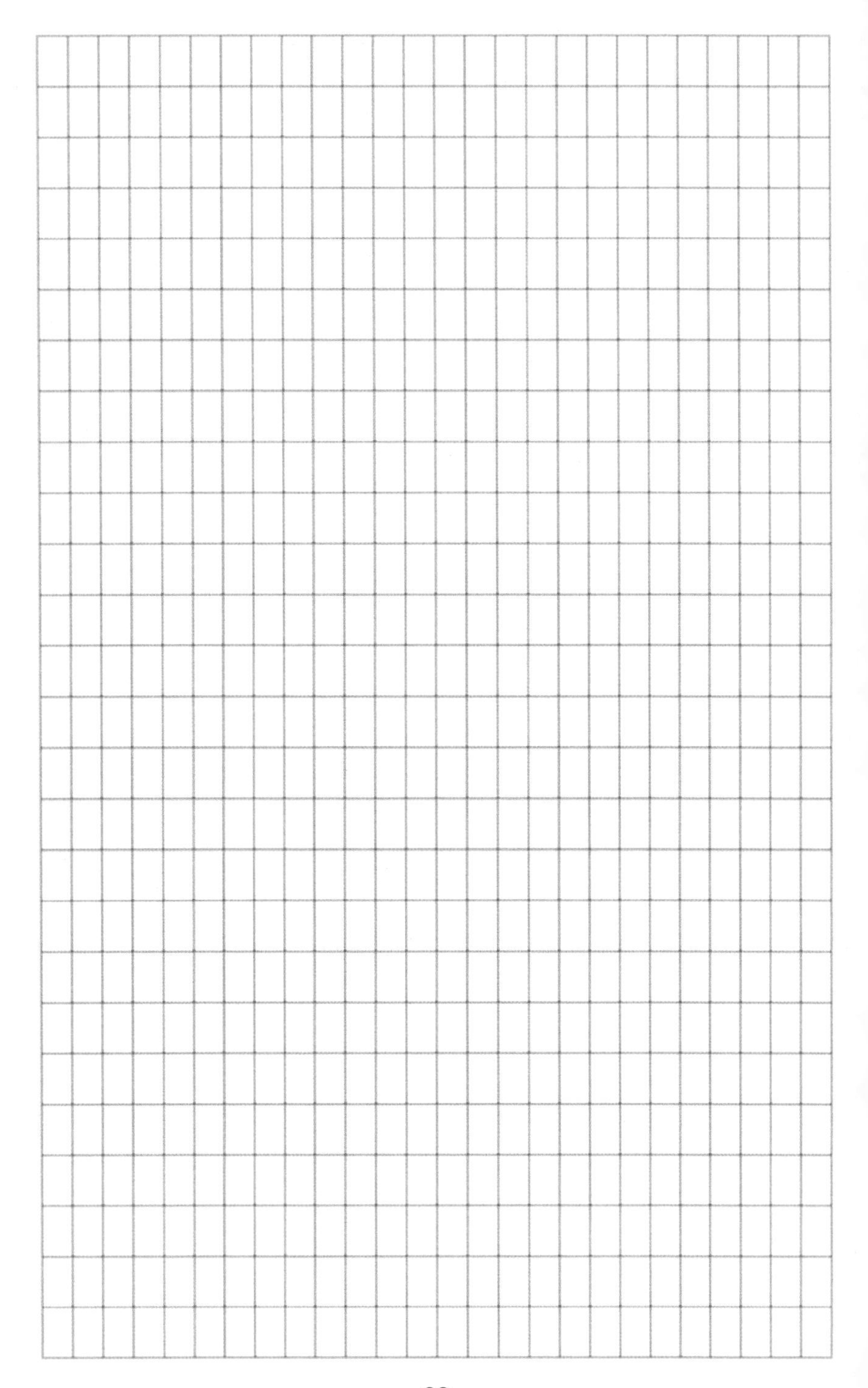

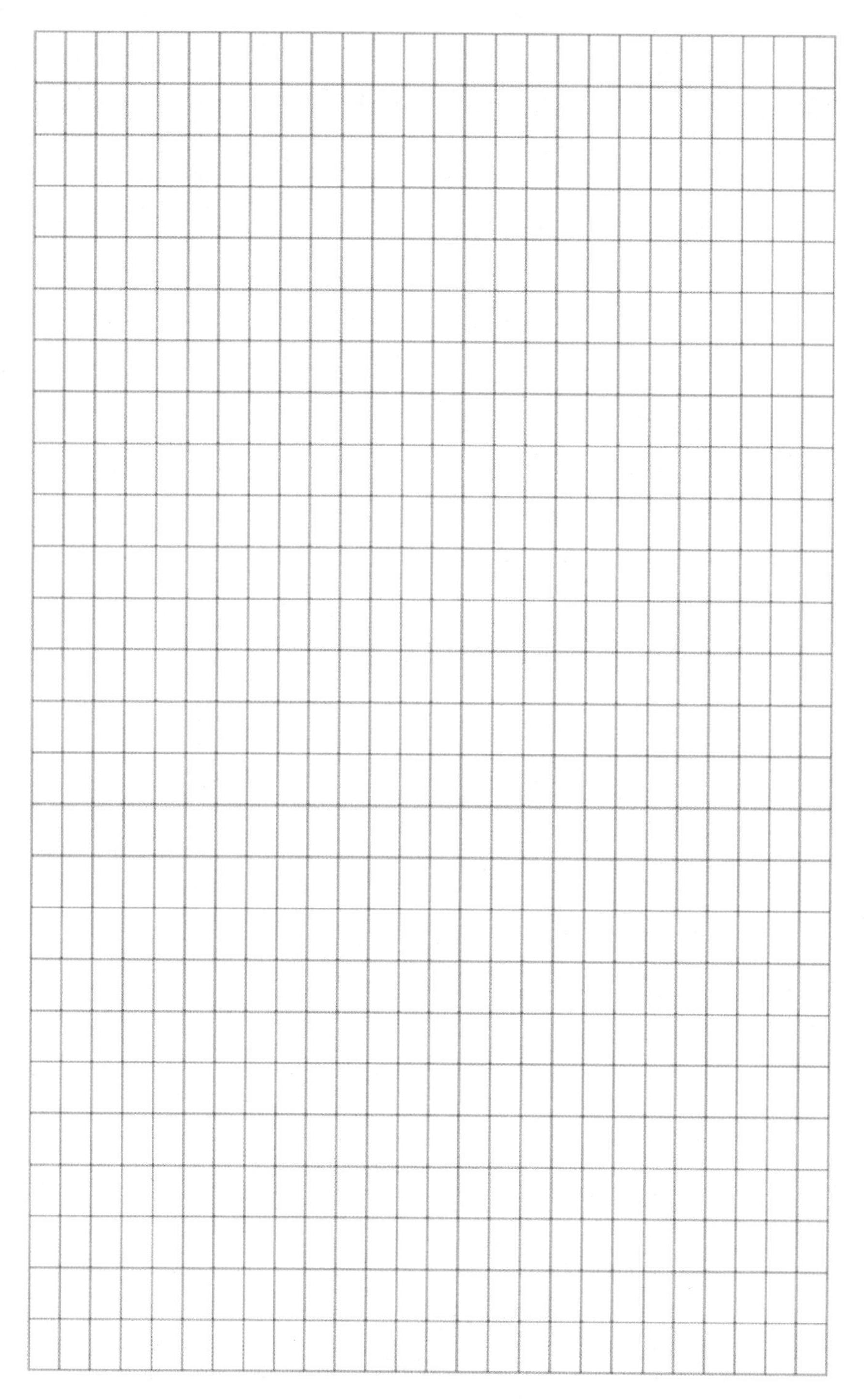

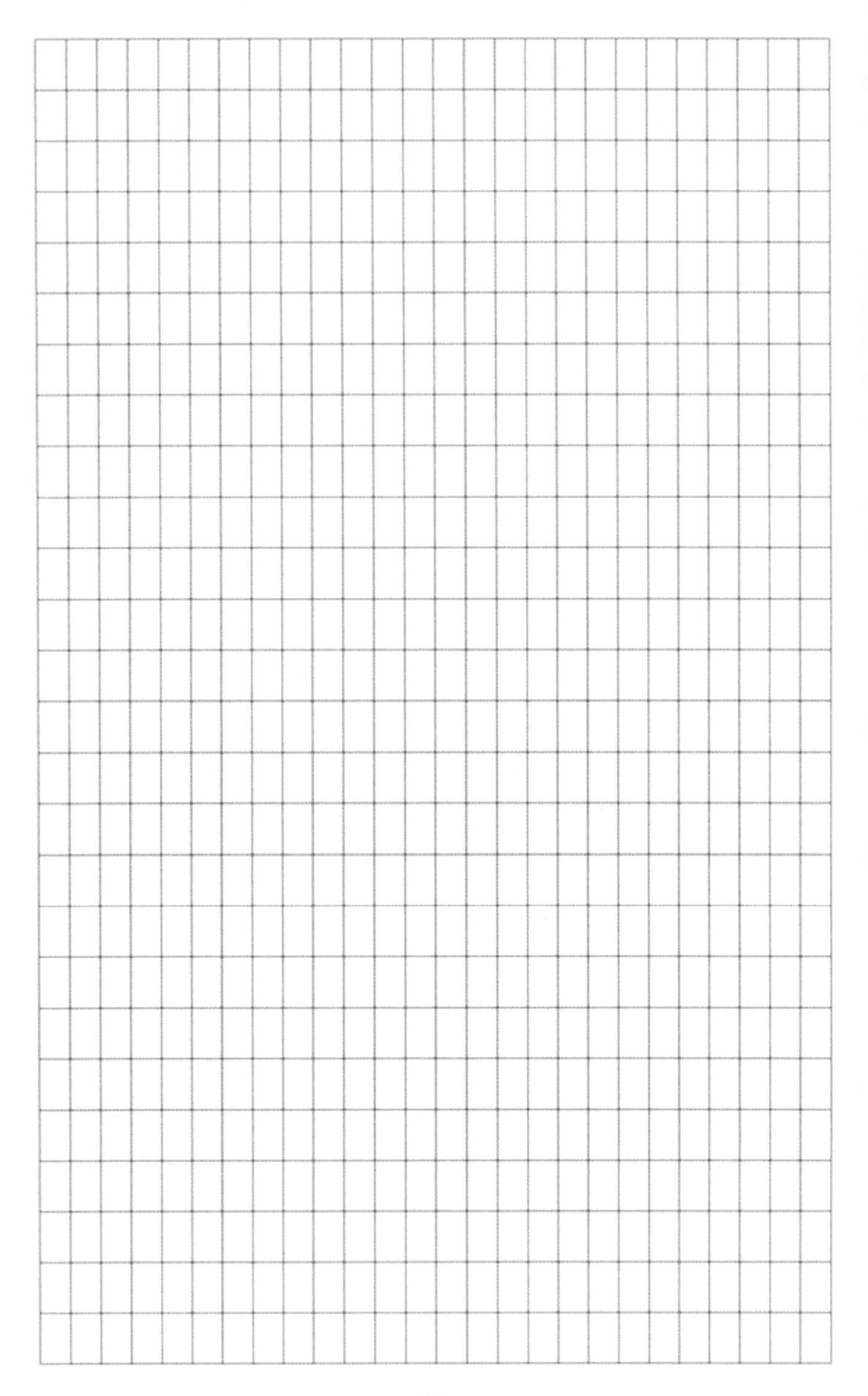

Made in the USA
Columbia, SC
03 March 2020